DOT-FTA-MA-26-7071-03-1
DOT-VNTSC-FTA-03-05

U.S. Department
of Transportation
**Federal Transit
Administration**

Clean Air Program

Design Guidelines for Bus Transit Systems Using Electric and Hybrid Electric Propulsion as an Alternative Fuel

March 2003
Final Report

OFFICE OF RESEARCH, DEMONSTRATION, AND INNOVATION

NOTICE

This document is disseminated under the sponsorship of the U.S. Department of Transportation in the interest of information exchange. The United States Government assumes no liability for its contents or use thereof.

NOTICE

The United States Government does not endorse products or manufacturers. Trade or manufacturers' names appear herein solely because they are considered essential to the objective of this report.

REPORT DOCUMENTATION PAGE			Form Approved OMB No. 0704-0188
Public reporting burden for this collection of information is estimated to average 1 hour per response, including the time for reviewing instructions searching existing data sources, gathering and maintaining the data needed, and completing and reviewing the collection of information. Send comments regarding this burden estimate or any other aspect of this collection of information, including suggestions for reducing this burden, to Washington Headquarters Services Directorate for Information Operations and Reports, 1215 Jefferson Davis Highway Suite 1204, Arlington, VA 22202-4302, and to the Office of Management and Budget, Paperwork Reduction Project (0704-0188), Washington, DC 20503.			
1. AGENCY USE ONLY (Leave blank)	2. REPORT DATE March 2003	3. REPORT TYPE AND DATES COVERED Final Report – March 2003	
4. TITLE AND SUBTITLE Design Guidelines for Bus Transit Systems Using Electric and Hybrid-Electric Propulsion as an Alternative Fuel		5. FUNDING NUMBERS U3077/TT39	
6. AUTHOR(S) William P. Chernicoff,* Thomas Balon**, and Phani Raj***.			
7. PERFORMING ORGANIZATION NAME(S) AND ADDRESS(ES) Volpe Center* MJ Bradley and Associates ** TMS Inc*** Cambridge, MA 02142 Manchester, NH Burlington, MA		8. PERFORMING ORGANIZATION REPORT NUMBER DOT-VNTSC-FTA-03-05	
9. SPONSORING/MONITORING AGENCY NAME(S) AND ADDRESS(ES) U.S. Department of Transportation Federal Transit Administration Office of Research Demonstration and Innovation		10. SPONSORING/MONITORING AGENCY REPORT NUMBER FTA-MA-26-7071-03-1	
11. SUPPLEMENTARY NOTES This work performed under contract to: U.S. Department of Transportation Volpe National Transportation Systems Center Cambridge, MA 02142			
12a. DISTRIBUTION/AVAILABILITY STATEMENT This document is available to the public through the National Technical Information Service, Springfield, VA 22161		12b. DISTRIBUTION CODE	
13. ABSTRACT (Maximum 200 words) The use of alternative fuels to power transit buses is steadily increasing. Several fuels, including Compressed Natural Gas (CNG), Liquefied Natural Gas (LNG), Liquefied Petroleum Gas (LPG), and Methanol/Ethanol, are already being used. At present, there are no available comprehensive facility guidelines to assist transit agencies contemplating converting from diesel to electric of hybrid electric propulsion. This document addresses that need. This guidelines document presents various facility and bus design issues that need to be considered to ensure safe operations when using electric or hybrid electric propulsion. Fueling facility, garaging facility, maintenance facility requirements and safety practices are indicated. Among the issues discussed are electric storage device properties, potential hazards, requirements for specified level of service, and applicable codes and standards. Critical fuel related safety issues in the design of the related systems on the bus are also discussed. A system safety assessment and hazard resolution process is also presented. This approach may be used to select design strategies which are economical, yet ensure a specified level of safety. This report forms part of a series of published by the U.S. DOT/FTA on the safe use of alternative fuels. Documents similar to this one in content have been published for CNG, Hydrogen, LPG, LNG, and Methanol/Ethanol.			
14. SUBJECT TERMS Electric propulsion, Hybrid-electric propulsion, electric drive, transit bus, transit facility design, system safety, alternative fuel		15. NUMBER OF PAGES 160	
		16. PRICE CODE	
17. SECURITY CLASSIFICATION OF REPORT Unclassified	18. SECURITY CLASSIFICATION OF THIS PAGE Unclassified	19. SECURITY CLASSIFICATION OF ABSTRACT Unclassified	20. LIMITATION OF ABSTRACT

NSN 7540-01-280-5500

Standard Form 298 (Rev. 2-89)
Prescr bed by ANSI Std. 239 18 298-102

Metric/English Conversion Factors

English to Metric

LENGTH (Approximate)
- 1 inch (in) = 2.5 centimeters (cm)
- 1 foot (ft) = 30 centimeters (cm)
- 1 yard (yd) = 0.9 meter (m)
- 1 mile (mi) = 1.6 kilometers (km)

AREA (Approximate)
- 1 square inch (sq in, in^2) = 6.5 square centimeters (cm^2)
- 1 square foot (sq ft, ft^2) = 0.09 square meter (m^2)
- 1 square yard (sq yd, yd^2) = 0.8 square meter (m^2)
- 1 square mile (sq mi, mi^2) = 2.6 square kilometers (km^2)
- 1 acre = 0.4 hectare (he) = 4,000 square meters (m^2)

MASS-WEIGHT (Approximate)
- 1 ounce (oz) = 28 grams (gm)
- 1 pound (lb) = 0.45 kilograms (kg)
- 1 short ton = 2,000 pounds (lb) = 0.9 tonne (t)

VOLUME (Approximate)
- 1 teaspoon (tsp) = 5 milliliters (ml)
- 1 tablespoon (tbsp) = 15 milliliters (ml)
- 1 fluid ounce (fl oz) = 30 milliliters (ml)
- 1 cup (c) = 0.24 liter (l)
- 1 pint (pt) = 0.47 liter (l)
- 1 quart (qt) = 0.96 liter (l)
- 1 gallon (gal) = 3.8 liters (l)
- 1 cubic foot (cu ft, ft^3) = 0.03 cubic meter (m^3)
- 1 cubic yard (cu yd, yd^3) = 0.76 cubic meter (m^3)

TEMPERATURE (Exact)
- [(x - 32) (5 / 9)] F = y C
- (x + 460) / 1.8 = y K

PRESSURE (Exact)
- 1 psi = 6.8948 k Pa

ENERGY & ENERGY DENSITY (Exact)
- 1 Btu = 1.05506 kJ
- 1 Btu/lb = 2.326 kJ/kg

Metric to English

LENGTH (Approximate)
- 1 millimeter (mm) = 0.04 inch (in)
- 1 centimeter (cm) = 0.4 inch (in)
- 1 meter (m) = 3.3 feet (ft)
- 1 meter (m) = 1.1 yards (yd)
- 1 kilometer (km) = 0.6 mile (mi)

AREA (Approximate)
- 1 square centimeter (cm^2) = 0.16 square inch (sq in, in^2)
- 1 square meter (m^2) = 1.2 square yards (sq yd, yd^2)
- 1 square kilometer (km^2) = 0.4 square mile (sq mi, mi^2)
- 10,000 square meters (m^2) = 1 hectare (he) = 2.5 acres

MASS-WEIGHT (Approximate)
- 1 gram (gm) = 0.036 ounce (oz)
- 1 kilogram (kg) = 2.2 pounds (lb)
- 1 tonne (t) = 1,000 kilograms (kg) = 1.1 short tons

VOLUME (Approximate)
- 1 milliliter (ml) = 0.03 fluid ounce (fl oz)
- 1 liter (l) = 2.1 pints (pt)
- 1 liter (l) = 1.06 quarts (qt)
- 1 liter (l) = 0.26 gallon (gal)
- 1 cubic meter (m^3) = 36 cubic feet (cu ft, ft^3)
- 1 cubic meter (m^3) = 13 cubic yards (cu yd, yd^3)

TEMPERATURE (Exact)
- [(9 / 5) y + 32] C = x F
- (y x 1.8 B 460) = x F

PRESSURE (Exact)
- 1 M Pa = 145.04 psi

ENERGY & ENERGY DENSITY (Exact)
- 1 MJ = 947.81 Btu
- 1 MJ/kg = 430 Btu/lb

QUICK FAHRENHEIT-CELSIUS TEMPERATURE CONVERSION

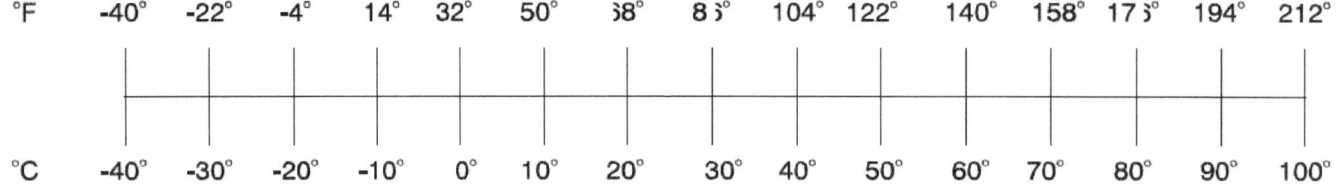

Acknowledgements

The work reported in this document was performed by Technology & Management Systems, Inc. and M.J. Bradley & Associates, Inc., under contract DTRS57-02-P-80173 from the U.S. Department of Transportation, John A. Volpe National Transportation Center ("Volpe Center") in Cambridge, Massachusetts. William Chernicoff was the Project Manager at the Volpe Center.

The Volpe Center wishes to thank the subcontractors, companies, and individuals that provided assistance to this project.

Companies and organizations involved in the review process included:

- Advanced Vehicle Systems, Inc.
- Aerovironment
- Advanced Transportation Technology Institute (Formerly ETVI)
- BAE SYSTEMS Controls, Inc.
- Federal Transit Administration TRI-20
- General Motors, Allison Transmission Division
- M.J. Bradley & Associates, Inc.
- MTA New York City Transit
- New York Power Authority
- Santa Barbara Electric Transit Institute
- Technology & Management Systems, Inc.
- Underwriters Laboratory

Table of Contents

Page

CHAPTER 1: Introduction 1

   1.1   BACKGROUND 1

   1.2   DISCLAIMER 1

   1.3   DOCUMENT OBJECTIVE 2
         1.3.1   Description of the Contents 2

CHAPTER 2: Overview of Battery Electric and Hybrid-Electric Bus Technologies. 3

   2.1   WHAT IS AN ELECTRIC BUS? 3

   2.2   WHAT IS A HYBRID-ELECTRIC BUS? 4

   2.3   HYBRID-ELECTRIC CONFIGURATIONS 5
         2.3.1   Series Hybrid 5
         2.3.2   Parallel Hybrid 5

   2.4   ELECTRIC AND HYBRID-ELECTRIC BUS COMPONENTS 6
         2.4.3   Battery Thermal Management System 11
         2.4.4   Replacement of the Battery Pack 13

CHAPTER 3: Electric and Hybrid-Electric Bus Safety Issues 14

   3.1   ELECTRIC SHOCK 15
         3.1.1   Enclose and Label High Voltage Parts 16
         3.1.2   Electrical Isolation 17
         3.1.3   Additional Recommendations 17
         3.1.4   Battery Box 18
         3.1.5   Automatic Disconnect Devices for Energy Storage Systems 18
         3.1.6   On-Board Charging 18
         3.1.7   Power Train and Control Systems 19

   3.2   HYDROGEN GASSING 19

   3.3   FIRE/TOXIC FUMES 20

   3.4   ELECTROLYTE (ACID) SPILLS 21

   3.5   VEHICLE (ELECTRICAL SYSTEM) MAINTENANCE ISSUES RELATED TO SAFETY 21

CHAPTER 4: Safety Issues in the Maintenance/Storage Facility 23

4.1   BATTERY OFF-LOADING & HANDLING 23

4.2   BATTERY STORAGE... 24

4.3   BATTERY CHARGING... 24

       4.3.1   Battery Charger Safety and Location 25

4.4   FACILITY FIRE DETECTION AND PROTECTION SYSTEMS 26

CHAPTER 5: Personnel Training 27

5.1   TRAINING OF TRANSIT VEHICLE OPERATORS 27

5.2   TRAINING OF MAINTENANCE PERSONNEL 28

5.3   EMERGENCY RESPONSE PERSONNEL TRAINING 29

CHAPTER 6: REFERENCES 31

CHAPTER 7: APPENDIX A: APPLICABLE REGULATIONS, CODES, STANDARDS & RESOURCES 32

REGULATIONS 32

CODES & STANDARDS 32

RESOURCES 34

List of Tables

		Page
2.1	COMPARISON OF HYBRID CONFIGURATIONS	6
2.2	ELECTRIC MOTOR COMPARISON	8
2.3	PERFORMANCE OF SOME ADVANCED ELECTRIC BUS BATTERY SYSTEMS	13
4.1	BATTERY CHARGER REQUIREMENTS	24

List of Figures

		Page
2.1	SCHEMATIC OF AN ELECTRIC VEHICLE	3
2.2	SERIES CONFIGURATION	5
2.3	PARALLEL CONFIGURATION	5
2.4	EFFECTS OF TEMPERATURE ON BATTERIES	12

ACRONYMS

AC	Alternate Current
APU	Auxiliary Power Unit
BMS	Battery Management System
CNG	Compressed Natural Gas
DC	Direct Current
EPRI	Electric Power Research Institute
FTA	Federal Transit Administration
FTP	Federal Transient Procedure
Li	Lithium
LNG	Liquid Natural Gas
NEC	National Electrical Code
NFPA	National Fire Protection Association
NiCd	Nickel Cadmium
NiMH	Nickel Metal Hydride
NOx	Nitrogen Oxides
NFPA	National Fire Protection Association
NYC	New York City
PM	Particulate Matter
SAE	Society of Automotive Engineers
UL	Underwriters Laboratories
US	United States
V	Volt
Volpe Center	US Department of Transportation, John A. Volpe National Transportation Center
Wh/kg	Watt-hours per kilogram

CHAPTER 1: Introduction

1.1 BACKGROUND

The Office of Research, Demonstration, and Innovation of the Federal Transit Administration (FTA) funded a number of research and demonstration projects involving the application of alternative fuel technologies to transit buses. FTA has also funded a number of electric and hybrid-electric bus demonstration projects. At present, there are approximately 220 electric buses, 90 hybrid-electric buses and trolleys, and 6 fuel cell buses operating in the US.[1] Electric and hybrid-electric buses offer transit agencies a way to reduce local emissions without potentially costly alternative fuel infrastructure costs.

FTA has published a set of five guideline reports for transit agencies planning to incorporate alternative fuel vehicles in their fleets. These documents are listed in Appendix A and are available for free to the public. FTA received positive feedback from the transit industry on the usefulness of these guidelines for operations, training, and bus procurement activities. Therefore the FTA initiated the development of this document for electric and hybrid-electric buses, similar in format and scope to the previous publications.

1.2 DISCLAIMER

While this guidance document was reviewed by a broad-based and representative group of individuals, none of the participating organizations were asked to, nor have they necessarily, endorsed or adopted the recommendations included in this guidance document. Nor does FTA endorse any company or individual who supported this effort or products that may be mentioned in the document.

This document is intended to be a guidebook on bus and facility design issues and SHOULD NOT be considered a specification manual or a substitute for existing local, state, or national codes and regulations. In addition, the reader should consider the following issues when reading this document.

- Every facility that is either modified or newly constructed should be in compliance with all local, state, and national codes and regulations.

- The information provided in this guidebook is by no means exhaustive on the subject of bus and facility design, personnel training or any other associated issues. The transit system should consult with knowledgeable engineers, consultants, fuel suppliers, design architectural and engineering firms, and the staff of the local authority having jurisdiction over the design of the facility consistent with local codes, regulations, and local conditions.

---

[1] Based on data from the Electric Vehicle Association of America's website http://www.evaa.org, 2002.

- The document references sections of national codes, recommended practices, and regulations. Such references to particular sections are NOT intended to convey the impression that only these sections apply. It is, however, intended to get the reader started or even directed to the appropriate sections in the standards or codes. It is recommended that the provisions of a currently adopted code or standard be reviewed thoroughly and in their entirety.

1.3 DOCUMENT OBJECTIVE

The principal objective of this document is to provide transit agencies with an overview of the technology, recommended safety specifications in bus design, and training for personnel that will enable them to understand the implications of purchasing, operating, and maintaining electric and hybrid-electric buses. In addition, the document is intended to provide basic information on electrical and operational safety for transit and non-transit personnel, such as emergency responders to an accident.

This document was not developed under a consensus format. Rather, the document highlights issues that transit agencies should be aware of and directs readers to review established industry standards, where available, for minimum specification requirements. Since many technology alternatives are being explored, it is difficult at this time to determine uniform standards that overlap the various technologies and configurations available in the electric and hybrid-electric bus market. This report should be used by transit systems to survey the various levels of safety and training options available and initiate discussions with manufacturers regarding these areas.

1.3.1 Description of the Contents

Chapter 2 outlines the principles of electric and hybrid-electric technologies and the major components as they apply to transit buses. This report does not include information on electric buses whose power source is external, such as an electric trolley bus. Chapter 3 examines safety issues that high voltage electric buses pose. Safety in garages and maintenance facilities is covered in Chapter 4. The types of training for personnel and emergency responders for transit agencies with electric and hybrid-electric buses are discussed in Chapter 5. Appendix A provides a list of rules, regulations, and standards that should be consulted to understand the requirements for electric and hybrid-electric buses and infrastructure. Additionally, this appendix includes recommended resources for more detailed information.

CHAPTER 2: Overview of Battery Electric and Hybrid-Electric Bus Technologies

To understand what makes the electric bus and hybrid-electric bus designs unique, it is helpful to compare them to a conventional bus. In a conventional bus design an internal combustion engine provides power. The engine is tied mechanically to the wheels via a transmission, allowing the vehicle to utilize the engine's energy directly. Fossil-fueled engines emit air pollution, such as nitrogen oxides (NOx) and particulate matter (PM). One of the goals of electric and hybrid-electric buses are to reduce or eliminate these exhaust pollutants.

2.1 WHAT IS AN ELECTRIC BUS?

The operating principle of a battery-powered electric vehicle is simple. An energy

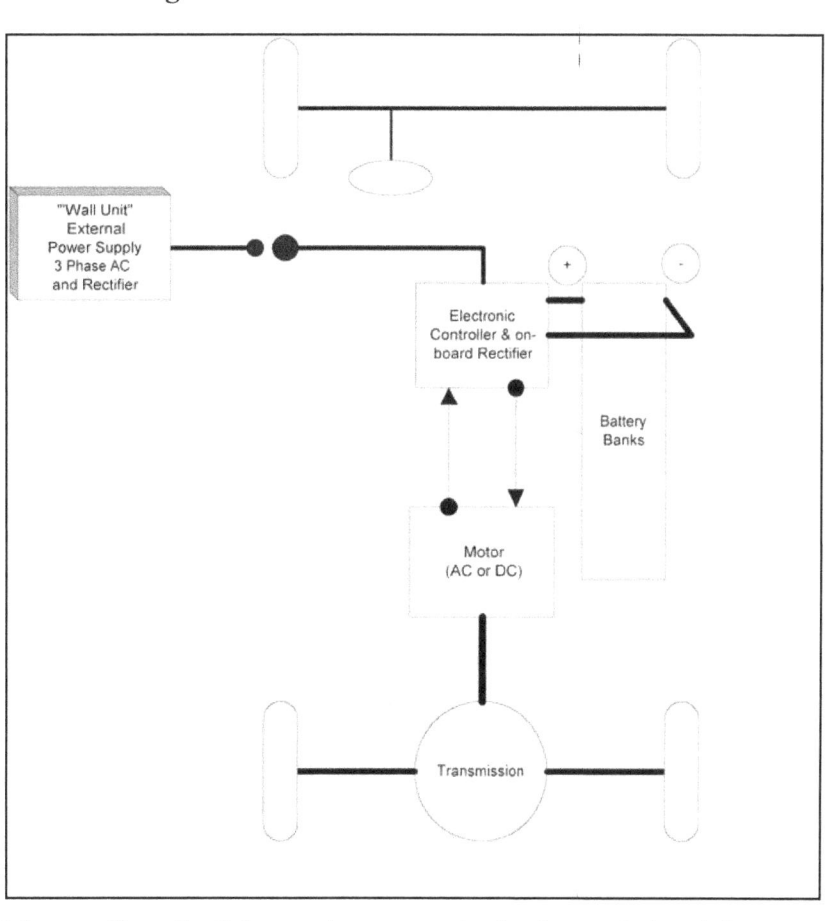

Figure 2.1: Schematic of an Electric Vehicle

storage device located on the vehicle supplies all of the motive energy in the form of electricity to a traction motor or motors. The rotary motion of the electric motor translates rotary motion to the vehicle wheels either by direct drive or through a mechanical transmission. The speed of the motor is controlled by an on-board electronic controller, which functions primarily based on the position of the accelerator pedal. The energy storage device, typically batteries, is recharged from an external electrical source when its charge is depleted. This mode of operation has an advantage over conventional and hybrid-electric buses, as it eliminates local air pollutants and engine noise. Figure 2.1 shows the important components of a battery-powered electric bus. These components are described in more detail later in the chapter.

> **Energy Storage Device:** A component or system of components that stores energy and for which its supply of energy is rechargeable by an electric motor system, an off-vehicle electric energy source, or both.

Most electric vehicles also recharge the energy storage device during braking by recovering part of the vehicle's kinetic energy.

Termed regenerative braking, the traction motor acts as a generator, and as a brake on the vehicle, with the electrical energy generated during braking fed into the energy storage device. There are a number of variations in the application of the above basic principle depending upon the vehicle size, duty cycle, and other technical and economic considerations.

> **Regenerative Braking:** Deceleration of the vehicle caused by operating an electric motor system, thereby returning energy from the vehicle propulsion system to the energy storage device or to operate auxiliaries.

2.2 WHAT IS A HYBRID-ELECTRIC BUS?

A hybrid-electric bus carries at least two sources of motive energy on board the vehicle. The non-electric source is typically referred to as an auxiliary power unit (APU)[2], which converts replenishable fuel into energy. Examples of APUs are internal combustion reciprocating engines, microturbines, or fuel cells. Depending upon the design of the hybrid-electric bus, an APU generates energy either continuously or intermittently. The energy is then used either to drive the wheels or stored for later use. The electrical energy produced by the generator may be used to charge the energy storage device or is directly fed to an electric motor to provide energy for motive power. The energy storage system may also be recharged by the energy recovered by regenerative braking. An electronic controller controls the "flow of current" from the batteries and/or the APU-generator set to the traction motor. The electronic controller is continually monitoring the energy storage device's state of charge to determine whether engine operation is needed to recharge the batteries independent of the driver signals. Introducing electrical energy as a means to provide a portion of the energy for vehicle motive power allows for a decrease in the total necessary energy and severity of transient operation required by the conventional reciprocating engine, therefore, reducing the amount of pollution emitted.

> **Auxiliary Power Unit:** Converts fuel into electrical energy. May take the form of an engine/generator, fuel cell, or turbine.

The principal hybrid-electric bus components include: (a) a drive motor, (b) a controller and inverter, (c) an energy storage device, (d) an APU, and (e) other auxiliary systems, such as air conditioning and lighting. An advantage of a hybrid-electric bus over a conventional bus is theoretically better fuel economy and lower exhaust emissions. A previous analysis of in-use data indicates that engines in series hybrid-electric vehicles exhibit less aggressive transient behavior with engine operating points closer together in a much smaller range.[3] This is because the energy storage system provides the extra power for acceleration and grade demands, allowing a less precise relationship between engine speed and wheel speed. Avoidance of certain engine operating regimes minimizes PM and NOx emissions and allows for the engine to be operated for optimum steady-state efficiency. Additional information on the increased efficiency and emission reductions can be found in the Northeast Advanced Vehicle Consortium's *Hybrid-Electric Drive Heavy-Duty Vehicle Testing Project* report.[4]

---

[2] In hybrid-electric buses the combustion engine may be called auxiliary even though it may provide a majority of the motive energy. If the vehicle cannot be operated in electric only mode for any significant distance, the APU may simply be referred to as the engine.

[3] Northeast Advanced Vehicle Consortium, M.J. Bradley & Associates, West Virginia University, *Hybrid Transit Bus Certification Workgroup*, NAVC, 0599-AVP009903, 2000. Available at www.navc.org.

[4] Available at http://www.mjbradley.com/reports htm

2.3 HYBRID-ELECTRIC CONFIGURATIONS

2.3.1 Series Hybrid

Hybrid-Electric technology is typically divided into two general types of drive configurations—series and parallel with a number of subcategories and even combinations of the two. In the series hybrid, similar to an electric bus, either a single electric motor drives the wheels through a mechanical transmission or an independent wheel motor drives each drive wheel. The electric motor(s) may draw energy from either the energy storage device or from the APU as determined by the controller. Figure 2.2 shows the series-hybrid system in a hybrid-electric bus.

Figure 2.2: Series Configuration

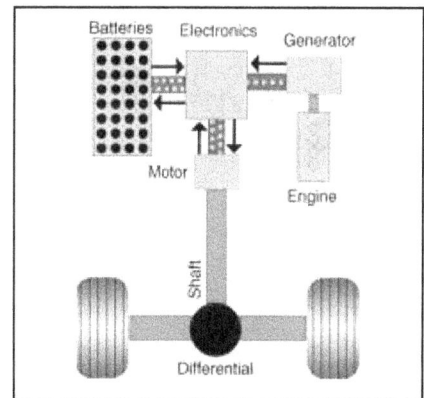

Source: Electric Transit Vehicle Institute

The two main variants of a series hybrid depend upon whether the APU or the battery dominates the system. In the engine-dominant hybrid, also referred to as charge sustaining, the engine provides a significant part of the drive power; that is, the energy is immediately utilized, minimizing efficiency losses that occur from energy storage. The size of the battery can therefore be small; however, the range in an all-electric mode would be relatively short. In the battery-dominant configuration, which can be charge sustaining or charge depleting, a greater distance can be traversed on the battery power alone and a larger quantity of the regenerated electrical energy can be stored. The disadvantage is that the size of the battery pack, and hence its weight may be significantly larger and there may be increased energy losses due to energy being routed through the batteries.

As noted, the determination of APU or energy storage system dominance is generally linked to whether the vehicle is charge sustaining or charge depleting. A pure electric bus is both charge depleting and battery dominant as it derives all of its motive energy from batteries and needs to be recharged from an external sources, such as the electricity grid. A charge depleting hybrid-electric bus is similar in that a majority of the energy is derived from the electricity grid and only minimal energy is supplied from the APU to extend vehicle range. At the other end of the spectrum, a charge-sustaining hybrid derives a majority of its energy from the APU and connection to the electricity grid is only used for battery conditioning if necessary.

Figure 2.3: Parallel Configuration

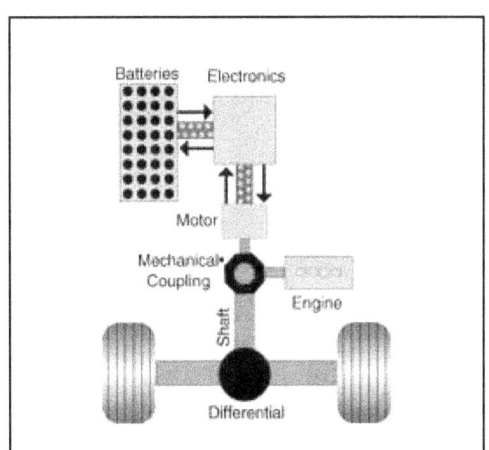

Source: Electric Transit Vehicle Institute

2.3.2 Parallel Hybrid

In a parallel-hybrid system the electric motor and the APU are both connected to the vehicle drive wheels. This system is shown schematically in Figure 2.3. The electric drive motor draws energy from the energy storage device, to supply additional tractive effort and also recovers energy from regenerative braking and supplies this energy back to the energy storage device. It is also possible to configure the drive system such that the APU can move the vehicle while simultaneously recharging the energy storage system. The parallel-hybrid configuration can be designed in either an engine-dominant or a battery-dominant subtype.

Each hybrid configuration has its own advantages and disadvantages, as listed in Table 2.1. The choice in hybrid-electric bus design is usually determined based on its intended duty application, such as central city urban versus long distance arterial service routes.

Table 2.1: Comparison of Hybrid Configurations

Type of Hybrid Configuration	Advantages	Disadvantages
Series	<ul><li>Allows APU to operate independently from the driver's commands for power, which reduces emissions</li><li>Fuel cell compatible</li><li>Energy efficient system when the vehicle is operated in stop and go modes</li><li>Transmission is eliminated</li><li>Electric only capable</li></ul>	<ul><li>Mechanical energy of the engine is converted into electrical energy and then reconverted to mechanical in the drive motor</li><li>Less suitable for high-speed highway cruising if equipped with a small APU</li></ul>
Parallel	<ul><li>The battery provides additional power during accelerations. Hence, the engine can be sized smaller than in conventional diesel buses for comparable accelerations</li><li>Direct mechanical drive path is more efficient in certain drive modes, particularly high-speed steady state, such as highway cruising</li></ul>	<ul><li>Does not typically facilitate the installation of a non-mechanical APU, such as a fuel cell</li><li>Less capable of capturing all available regenerative braking energy when small battery packs are used in engine dominant designs</li></ul>

2.4 ELECTRIC AND HYBRID-ELECTRIC BUS COMPONENTS

Auxiliary Power Units: APUs used in hybrid-electric buses are available in a number of configurations including reciprocating internal combustion engines, fuel cells and microturbines, and with different fuels, such as diesel, gasoline and compressed natural gas (CNG), liquid natural gas (LNG) and propane. The choice of the APU affects the performance of the bus, overall efficiency and emissions as is true with a conventional bus. The following describes four types of APUs.

Internal Combustion Engines: Engines utilized in hybrids can be the same engines used in conventional buses. However, they tend to be smaller, because the bus does not rely entirely on the engine for peak power output at the axles. Instead, the engine is typically sized for the average bus power demand, not peak power demand since the energy storage device provides supplementary power. The advantage of hybrids is that for the same output, a smaller engine operating at a higher percentage output can be more efficient than a larger engine operated at a lower percentage output due to internal engine losses. Engines in hybrid configurations also operate over a narrower range of load and speed combinations compared to engines in conventional buses. The power rating for diesel engines for 40 foot buses operating in a hybrid configuration range can be as small as 150 horsepower, compared to 250 - 275 horsepower in the same size conventional bus. In a series application the engine can be determined on its power, not its torque, which may further increase the system fuel efficiency and fuel options.

One of the obstacles that hybrid-electric bus manufacturers face is that the engines for heavy-duty transit buses must be certified for both emissions and durability using the Federal Transient Procedure (FTP). However, the smaller sized engines that are optimized for series hybrid-electric buses are not typically utilized or certified for use in conventional buses. Thus, even though they meet the emission requirements in a hybrid configuration, the engine may not by itself meet the durability requirements.

Microturbines: While a microturbine could not be utilized in a conventional bus, it is suitable for a hybrid bus service because its power output can be directly fed to an electric generator. Microturbines have the advantage of few moving parts, which reduces maintenance requirements and noise, and they tend to be lighter than diesel engines of similar power rating. Peak efficiency is maintained at near steady-state operating conditions and they can achieve more complete combustion, which means fewer emissions. However, they typically achieve lower overall fuel efficiency than other APUs due to limited overall effective combustion ratio. The high microturbine operating temperature also requires the installation of a heat recovery system. The special materials required to withstand the high temperatures and the precision in manufacturing of parts causes microturbines to be more expensive than comparable power diesel engines, which may be offset with the microturbine's lower maintenance requirements.

Fuel cells: Fuel cells generate direct current (DC) electricity from the chemical reaction between hydrogen and oxygen ions in a cell, facilitated by a catalyst. Fuel cells are attractive as APUs because there are relatively few moving parts, which reduce maintenance costs, and the only by-products are water and thermal energy when utilizing a pure hydrogen fuel. In theory, fuel cells also have a high efficiency for converting chemical energy into electricity. However, energy consumption for producing pure hydrogen fuel can reduce the overall energy efficiency when considered from a well to wheels basis.

Traction Motors: Two primary types of electric motors can be used in electric vehicles, DC motors and alternate current (AC) motors. On a power comparative basis, an AC motor generally exhibits higher efficiency, has a favorable power to size/weight ratio, is less expensive and generates regenerative braking energy more efficiently than a DC motor. However, AC motors require an inverter and more expensive controller, increasing the associated cost. An electric vehicle power train design based on a DC motor may be slightly less efficient overall and the DC motors themselves are more expensive. However, the controllers for DC motors are generally less expensive making the total cost compare between the two types of motors.

A motor is typically chosen based on specific considerations of the vehicle application, such as the road-load profile, maximum speed of the duty cycle and the maximum grade. Where a regenerative braking system is employed, the motor also doubles as a generator producing electricity during vehicle braking. AC motors are more common in buses than DC motors. Table 2.2 shows the characteristics of the two types of motors used in electric buses and hybrid-electric buses.

Table 2.2: Electric Motor Comparison

AC Motor	DC Motor
Single-speed transmission	Multi-speed transmission
Less expensive	More expensive
95% efficiency at full load	85 – 95% efficiency at full load
Motor/controller/inverter more expensive	Motor/controller less expensive

Source: US Department of Energy, "Electric Bus Power Systems," National Alternative Fuels, 2002.

Even efficient motors lose some energy input as heat. If this heat is not dissipated, the motor will overheat and fail or operate at a reduced efficiency. Either air-cooling or water/fluid cooling can be employed depending on the design, size and power rating and operating conditions of the motor. While water/fluid cooling is more efficient, potential disadvantages are increased system complexity and maintenance.

Electric drive motors are connected to the vehicle wheels either directly, referred to as wheel motors, or through a transmission and ring and pinion/differential assembly. Wheel motors are more efficient both in drive cycle and in the regenerative cycle by eliminating the losses in the mechanical transmission and the differential. The use of wheel motors has an added benefit of accommodating the "full" low floor bus design. However, wheel motors are expensive, have somewhat lower reliability to date, and the cooling systems tend to be more elaborate and complex in design. There is also a concern about heavy un-sprung mass that must be accounted for in suspension design. At least two wheel motors are needed to drive a vehicle, whereas drives through transmission and differential systems may be designed with a single motor. Improvements in cost, durability, and efficiency in both AC and DC motor technology are constantly occurring.

Controller and Inverter/Rectifier: The electronic controller regulates the amount of energy, (DC power in the case of batteries), that is transferred or converted to AC power by the inverter (in AC motors) for acceleration. It also ensures that voltage is maintained within the specifications required for operating the motor. An electronic controller can also recover electrical energy by switching the motor to a generator in order to capture the vehicle's kinetic energy via regenerative braking. The controller also ensures that the regenerative current does not overcharge a battery.

Inverters are also used on board pure electric vehicles when the drive motor is an AC motor; the DC voltage of the battery is converted into AC for powering the motor. In such a design, the regenerative breaking energy generated is in the form of AC current and is "rectified" or converted to DC by the "rectifier." The batteries always require DC current to recharge.

A loss of energy as heat is inevitable in both inversion, as well as rectification. Hence, for optimal performance, the inverter and rectifier must be either designed to withstand high temperatures or be provided with active cooling by a fan or liquid cooling. The conversion efficiency of these devices ranges from 87 percent to 96 percent.

Additionally, in grid-connected electric vehicles, the controller acts as an interface between the external charger and the on-board battery pack to ensure the optimal strategy for charging the batteries. It monitors the voltage and temperature of the battery pack and maintains the temperature within specified limits to ensure optimal battery performance. Monitoring, in more advanced controllers, can also include tracking of individual cell voltages in the battery to ensure a balanced battery operation.

Energy Storage Devices: Energy storage devices provide all of the energy in electric buses, and are also necessary in hybrid-electric buses to supplement the APU energy when there is a high demand (e.g., acceleration from stop, speed acceleration, climbing an up-hill gradient) and to recover and store the energy generated during deceleration (e.g., braking, down-hill coasting). The major challenge when choosing an energy storage device is to minimize size and weight, while maintaining or improving vehicle performance and efficiency. Batteries represent the single most reliable and proven technology among the energy storage devices commercially available in the transportation industry, although they are not the most efficient manner for energy storage. Considerable research is underway to utilize alternate types, including flywheels and ultracapcitors.

> **Energy Storage Devices**
>
> **Battery:** A device that stores chemical energy and releases electrical energy.
>
> **Capacitor:** A device that stores energy electrostatically and releases electrical energy.
>
> **Electromechanical Flywheel:** A device that stores rotational kinetic energy and can release that kinetic energy to an electric motor system, thereby producing electrical energy.

A battery system used in a vehicle consists of several individual batteries or modules connected in series to provide the required voltage for the vehicle. Each battery further consists of individual cells connected serially; any imbalance in the cell voltages may result in degradation in the overall battery performance. Battery technology is still evolving with a number of chemical combinations. The most common types used in transportation include lead-acid, nickel-metal hydride (NiMH), nickel cadmium (NiCd) and lithium (Li). Each one of these battery types has specific characteristics that can improve or decrease the performance of an electric vehicle. Lead-acid batteries have been utilized traditionally because of their proven reliability and relatively low cost. More recently a trend has been leaning towards lithium battery chemistries, which have been found to be more energy efficient with longer lifespans, but with a higher associated cost. In practical terms, the choice of a battery often comes down to a trade-off between performance and cost.

Below are some important issues related to the selection of batteries.

1. Energy Density: Battery weight can be a significant part of the overall weight of a bus. For example, a lead-acid battery pack with its corresponding bus packaging can weigh as much as 4,000 to 5,000 pounds. Thus a battery system with a high gravimetric energy density, rated in terms of watt-hours per kilogram (Wh/kg), will provide for better performance and

range. Lead-acid batteries have a relatively low energy density (30 to 40 Wh/kg), while NiMH and NiCd have a higher range (50 to 60 Wh/kg).

2. Battery Lifecycle: Batteries are a high replacement cost item and factor greatly into the reliability and operating cost of the bus. Low cost batteries usually need to be replaced more often. It is prudent to look at the total lifecycle cost of the batteries and not simply the price of the pack or individual module.

3. Recharge Time: Once battery packs are depleted of energy, the recharge time for lead-acid batteries can be lengthy (six to eight hours), although new charging technologies may be able to decrease this time. Other types of battery chemistry, such as NiCd and NiMH can sustain higher voltage recharging methods, which can reduce the recharge time to less than an hour.

4. Power Density: The power of a cell is the ability to discharge and accept energy at a given rate. This parameter indicates how rapidly the cell can be discharged and how much power is generated and is expressed in units of watts per kilogram. This characteristic helps determine the magnitude of acceleration of a vehicle. The higher the power density of a battery is, the higher the acceleration that can be achieved. However, the higher the power density is, the quicker the battery may be discharged and, therefore, the shorter the duration over which the battery is effective. In general, batteries with higher energy densities exhibit significant voltage and capacity drops at higher discharge rates and therefore have a lower power density.

5. Charge/Discharge Efficiencies: The efficiency of a battery is defined as the ratio of the energy delivered by a battery during discharge to the total energy required to restore it to a full state-of-charge. The battery efficiency decreases with usage and is a function of the number of charge-discharge cycles.

6. Maintenance: Both flooded lead-acid and NiCd batteries need to be refilled regularly to replenish the water that is lost to electrolysis during the battery charge and discharge cycles. Flooded lead-acid batteries need distilled water replenishing about twice per week and NiCd about once every 10 days.

7. Cost: Due to the use of unique metals, production volumes, or the status of the battery technology, the price range of batteries varies widely. Lithium batteries are the most expensive, while lead-acid batteries are at the lower end of the price range.

8. Recycling: Many manufacturers provide detailed information on the disposal or recycling of batteries. These protocols should be followed to ensure that hazardous materials in batteries, such as lead, cadmium, and corrosives are not released into the environment when the useful life of the battery is completed. Fluids used for the cooling of components are generally similar to other vehicle fluids, such as water, oil, and glycol, and should be disposed of properly.

9. <u>Interchangeability</u>: When replacing batteries of one manufacturer or model with that of another, care should be exercised to ensure that the battery specifications are the same and that they have the same ratings in power, power density, size and voltage. Battery switching may require checking with the original equipment manufacturer to ensure compatibility with the warranty. Generally, all batteries in a pack should be of the same type, manufacturer and condition.

10. <u>Safety</u>: The precautions to be taken in placement of the battery pack and personnel training depend on the type of battery. For example, lead-acid and NiMH batteries can potentially release hydrogen when overheated. Hence, the storage box needs to be well ventilated and the temperature needs to be monitored. Also, there is the potential for electrolyte leaks from batteries.

11. <u>Application Drive Cycle</u>: The magnitude of accelerations and decelerations experienced by a bus and hence, the rates of energy drawn from and retained to the batteries greatly affects the overall range and efficiency performance of the bus. The distance to which a bus can be driven on a fully charged battery pack is dependent upon a driver's ability to operate the bus; the greater the accelerations and decelerations, the lower the range on a single charge. This is because: (1) fast accelerations draw energy from the battery more quickly, resulting in lower "effective" battery capacity, and (2) quick decelerations are not efficient for capturing regenerated energy due to limitations on acceptance by the battery pack.

    While the range of a typical hybrid-electric bus may not be significantly affected by battery capacity, battery efficiency will impact fuel economy and emissions. In an electric bus with a slow operating cycle, a lead-acid battery may be a better choice, while a transit system with a fast transient cycle application might choose NiMH.

12. <u>Warranty</u>: A transit agency should check with the battery manufacturer that the transit system's battery charger and charging procedure is appropriate, and will not void the battery warranty. For example, some battery manufacturers are hesitant about fast charging battery packs.

Table 2.3 provides additional information on different types of batteries. A good source of information on batteries is the United States Advanced Battery Consortium, whose contact information is listed in Appendix A.

2.4.3 *Battery Thermal Management System*

Batteries also require a thermal management system because the discharge and recharge processes result in a net production of heat. If this heat is not dissipated quickly, the battery temperature can increase substantially resulting in accelerated self-discharge, degraded battery cycle life or battery failure. The efficiencies of a battery to accept or discharge current are also dependent on battery temperature. For example, lead-acid batteries are negatively affected by the cold, but other battery types may have issues with hot temperatures. This is also true of conventional buses, which see a degradation in efficiency in cold temperatures.

Table 2.3: Performance of Some Advanced Electric Bus Battery Systems

Battery Mfgr/Types	Energy Density (Wh/kg)	Power Density (W/kg)	Life Cycles per battery
Advanced Lead Acid	48	150	800
GM Ovonic NiMH	70	220	>600
SAFT NiMH	70	150	1500
SAFT Lithium Ion	120	230	600
Lithium Polymer	150	350	<600
Zebra Sodium-Nickel Chloride	86	150	<1000
USABC Short-Term Goals	*86*	*150*	*600*
USABC Long-Term Goals	*200*	*400*	*1000*

Source: US Department Of Energy, "Electric Bus Batteries," DOE Fields Program, 2002, available at http://ev.inel.gov/fop/general_info/battery.html

Figure 2.4 shows lead-acid and NiMH battery behaviors at lower than normal temperatures. It is recommended that batteries used on electric and hybrid-electric vehicles have a thermal management system, which regulates the temperature in the battery pack within tolerable limits, which is important for warranty and life expectancy.

Figure 2.4: Effects of Temperature on Batteries (Normalized to 21°C)

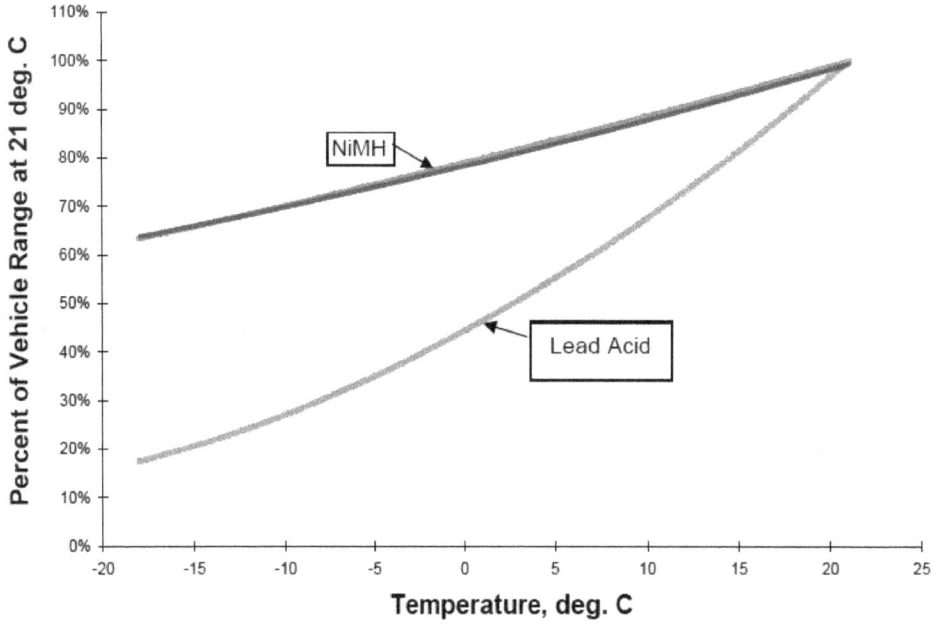

Source: EVermont, Advanced Battery Management and Technology Project, NAVC1096-PF009524, August 1999.

It is also recommended that the battery pack have a battery management system (BMS). This system consists of a microprocessor that monitors energy, as well as temperature, individual cell or module voltages, and total pack voltage. The BMS can adjust the control strategy algorithms to maintain the batteries at uniform state of charge and optimal temperatures.

2.4.4 Replacement of the Battery Pack

Eventually an electric bus or hybrid-electric bus operator will need to replace the bus battery pack or modules within the pack. If the electric bus or hybrid-electric bus experiences reduced battery range or voltage fluctuations before the predicted end of life and while operating within normal parameters, such as temperature, then individual module failures, or even poor connections, may be the cause of the malfunction. Swapping individual modules will require that the new replacement modules be at the same approximate state of charge as the rest of the pack or that the bus is equipped with a battery management system. Analyzing an entire battery pack is labor intensive, so current wisdom dictates that a malfunctioning battery pack more than halfway through its life cycle should be replaced and recycled.

Rectifier/Charger (external to vehicle): In general, 3 phase, 240/480 Volt (V) AC current is supplied to the batteries from a wall transformer/rectifier unit, which converts the AC electricity from the electric utility grid to DC. Normal electric bus battery pack voltage ranges from 240 to 360 VDC, although electric vehicles can have voltages between 100 V and 600 V (nominal). Battery charger designs come in a number of different charging strategies based on the way they control the recharge rate, such as current charge, constant voltage, or fast charge followed by cell voltage balancing.

Total battery recharge time is dependent on voltage, current, and the charging algorithm. Batteries can be damaged from the heat produced from charging at a high current, or from overcharging. Fast chargers have advanced in recent years, generating less battery heating and, therefore, lowering the risk of battery overheating. Slow chargers are less expensive and their low current rate minimizes potential damage to the battery cells. Most charging systems connect the supply and the bus using a conductive cable, although advances are also being made with inductive charging.

A manufacturer-supplied charging protocol generally involves a rapid charge period followed by a slower "cell voltage balancing" charge. In fleet operations, electric bus batteries are typically charged overnight when the electricity rates are lower. In several areas, such as California, New York, and Arizona, the demand charge for using electricity during peak daytime hours may be ten to a hundred times more expensive than during the night hours. Given high demand charges it may be advantageous to charge as slowly as allowable for a given application and to minimize opportunity charging during the day. Load leveling chargers may also be used that draw a steady current from the electricity grid, store that energy and deliver that energy to the vehicle as a fast charge.

Transit systems that do not experience daytime peak electricity charges may be interested in "opportunity charging" of batteries during service time. Other transit systems buy multiple battery packs to swap in freshly charged battery packs during a 15-minute break when battery energy is running low.

CHAPTER 3: Electric and Hybrid-Electric Bus Safety Issues

The introduction of electric drive and energy storage systems into buses adds several new safety considerations beyond the risks of conventional buses. A major difference is that both the electric bus and hybrid-electric bus have high voltage systems, whereas conventional buses typically have only a 24 V battery system. In addition to design considerations, electrical power systems have profound implications for maintenance operations. The potential additional hazards that electric bus and hybrid-electric bus generate are:

- Electric Shock
- Fire/Toxic Fumes
- Hydrogen Gassing
- Electrolyte (Acid) Spills

In general, most transit personnel are unfamiliar with high voltage safety procedures or they have not undergone the necessary training, and may also lack experience in servicing high-voltage systems and electric drive technology. Transit agencies should include certain safety precautions in bus specifications to minimize risks to passengers, drivers and maintenance personnel. Training for emergency response personnel is also a very important step when incorporating advanced vehicles into fleets. Typically this means going beyond the basic training that manufacturers may offer. Several organizations, such as Mid-Del Technology Center and National Alternative Fuels Training Consortium offer advanced training, and other transit agencies with these types of vehicles may offer the opportunity for personnel to train with them.

> **Electric and Hybrid-Electric Training Centers**
>
> Mid-Del Technology Center, Oklahoma, 405.672.6665
> www.evtraining.com
>
> National Alternative Fuels Training Consortium, West Virginia/Nationwide
> 304.293.7882
> www.naftp.nrcce.wvu.edu

The general safety requirements that transit buses, including alternative fueled ones, must meet are the Federal Motor Vehicle Safety Standards (49 CFR 571) issued by the National Highway Traffic Administration. It is also recommended that transit agencies review the American Public Transit Association's *Standard Bus Procurement Guidelines*, which provides crash-worthiness criteria advice when purchasing buses.[5] Other equipment codes and specifications are available from Underwriters Laboratory (UL) and Society of Automotive Engineers (SAE).[6]

In addition to federal regulations, the National Fire Protection Association establishes the National Electric Code (NEC) standards for electrical construction and operation. Articles 511 and 625 apply to the infrastructure (garages and charge equipment) necessary for electric and hybrid-electric fleets. Additionally, states and local cities/counties administer regulations related

---

[5] Available at http://www.apta.com/services/procurement/
[6] More information about SAE and UL are provided in Appendix A.

to electric vehicles and electricity infrastructure, which tend to be consistent with NEC. Transit agencies should check with the state and local building code commission, and state and local fire marshall office for other applicable regulations.

3.1 ELECTRIC SHOCK

The single most important difference between a conventionally fueled bus and an electric bus or a hybrid-electric bus is that, in the latter, there are high voltage circuits that may be active even when the bus is not operating. The effective operating voltage range for electric buses is about 300 V to more than 600 V, but can be as high as 800 V. Before any of the components in an electrical circuit are probed or maintained, personnel must be properly trained to avoid a high voltage shock.

Generally, the electrical systems in electric buses and hybrid-electric buses contain: (1) a low voltage (12 V - 24 V) system for such accessories as the lights, wiper, horn, sensors and instruments, and (2) a high voltage system (50 – 800 V DC and AC) including the traction motor, traction battery pack, rectifying and inverter systems and charging. The low voltage system should be independent of the high voltage system, so that in the event of a high voltage failure, emergency lighting and other accessory devices still operate.

A shock hazard can occur by:

- Contact with energized components;

- Contact with parts energized by insulation failure or short circuit; or

- Discharge of stored energy in the system even after primary power has been shut off.[7]

The risk of electric shock can be mitigated through proper engineering, labeling, and safe maintenance practices. In addition to complying with federal and state regulations, the design of an electric or hybrid-electric bus should meet the following criteria:

- Electrical systems and equipment should conform to the appropriate SAE standards or SAE recommended practices.

- UL standards should be consulted for off-board charging systems, as well as on-board appliances that connect to off-board power sources.

- A master disconnect switch is recommended, located at the driver's position. While a boarding passenger should not normally be able to reach this switch, someone should be able to reach the switch from the vestibule should the bus operator be disabled or unconscious. It would also be advantageous for emergency service personnel to be able to reach this switch while standing on the ground outside the driver's window.

---

[7] Controllers and inverters may utilize capacitors internally, which may store a limited amount of energy at substantial voltage.

- Fail-safe ground fault detection should be utilized. During normal operation ground path resistance should be less than 10 milliohms. The warning should be visible to both the driver and in the engine and battery compartment for maintenance personnel. A staged warning and shutdown is recommended. The first stage should annunciate the ground fault failure so that the vehicle can be serviced. The second stage should indicate an immediately dangerous condition and should precipitate a shutdown.

- If the bus can operate with zero, or minimal engine and/or equipment noise, then the operator should consider equipping the bus with an external, audible warning device to alert pedestrians and other drivers that a bus is turning or approaching.

- The bus should not be MOVABLE when the charging door is open. It is also preferable to detect an open charging door than to check for current at the charging port, since the cable may be attached, but not energized.

- The bus should be clearly marked "ELECTRIC POWERED" on the exterior.

3.1.1 Enclose and Label High Voltage Parts

Refer to SAE J1673 for specific minimum requirements. The bullets listed below highlight important areas from this recommended practice. Note that SAE J1673 references UL specifications, as well as other SAE recommended practices.

- There should be no high voltage areas within the passenger compartment and no exposed conductors, terminals, contact blocks or devices that create the potential for personnel to be exposed to greater than 50 V.

- All high voltage wiring and equipment should be shielded from "casual contact" by the removal of at least one bolt, screw or latch.

- All visible equipment or conductors operating at a voltage greater than 50 V should be identified with a HIGH VOLTAGE marking and warn of precautions to take if compartments are opened.

- Any door, cover or other panel that allows immediate access to a high voltage area or high voltage connectors shall be marked clearly with a warning, and the voltage. For example "DANGER -- 600 VOLTS DC." To avoid this warning on the exterior of buses, the vendor could add an interior, screw fastened panel to the high voltage equipment bay.

- High voltage wiring should be permanently identified with the use of orange color per SAE specifications.

Transit agencies should also request non-proprietary manuals for parts, service, operation and maintenance, interconnection wiring diagrams and schematics. The schematic should show all major components, switches, fuses and circuit breakers.

3.1.2 Electrical Isolation

Another important safety issue of concern when using high voltage electrical systems on vehicles is the potential for occurrence of fires caused by short circuit (from cables whose insulation is damaged) or from overheating due to excessive currents through the wires or the machinery. See SAE J1766 for specific recommendations.

- High voltage circuits (> 50 V) should be isolated from a vehicle chassis. High voltage and low voltage circuits should be physically separated to the extent practical.

- Wiring and electrical equipment located under the chassis should be protected against water, corrosion (from salt or moisture), heat and mechanical damage. Vibration and fatigue damage to wiring should be minimized by providing loops, elastomer grommets and strain relief, minimizing wire slack and avoiding sharp metal edges.

- Proper grounding must supply sufficient amperage for normal circuit operation and a proper ground return path for over current protection device operation.

- Circuit breakers and/or fuses should be provided to affect electrical isolation of components and systems in case of short circuits. See SAE J1763 for further information.

- Incorporate automatic electrical shut down and battery isolation in case of excessive current draw and/or short circuit. See SAE J1766 for specific ohm per volt thresholds that indicate a loss of isolation.

- Provide warning notice within the battery compartment and on the outside (for the benefit of emergency response personnel) NOT to pour water on the battery equipment in case of fire.

- It is recommended that the bus carry in an easily accessible place for the driver, a fire extinguisher with a UL rating of 10 A-B-C.

3.1.3 Additional Recommendations

Energy storage devices should be provided with cut out switch(es) to provide electrical isolation of the high voltage system power from all other electrical systems in the vehicle. The design of these switches should provide for their remote operation or hand operation. The cut out switch(es) should be installed at a location that is easily accessible to emergency responders. They should be labeled and easily understood by an individual unfamiliar with electric vehicles, and optionally include physical lock-out/tag-out features for maintenance. The electrical system should conform to SAE standards for wiring (J1654 and J1673) and connectors (J1742).

- When the master disconnect switch is in the "Off" or "Park" position, and the charging cord is not connected, there should be no voltages higher than 50 V present outside of the battery enclosures. If there are service outlets on the bus that provide power for service or other uses, these outlets should also be disabled.

- If the 12/24 V DC auxiliary battery is removed from the bus, all high voltage should be isolated within the battery boxes, regardless of the position of the master switch.

- Provide a means of informing the operator of the loss of high voltage ground isolation by proper annunciation on the dashboard with visual and optionally audible signals in a phased warning and shutdown.

3.1.4 Battery Box

- Battery box enclosures should be properly grounded and considered part of chassis ground for both electrical and structural integrity purposes.

- Insulation used in the battery box should not be hydrophilic and/or capable of absorbing moisture.

- Unless a battery compartment is hermetically sealed, designs should assume that moisture and dirt accumulation can occur.

3.1.5 Automatic Disconnect Devices for Energy Storage Systems

Buses should have an automatic method to disconnect an energized system from the electricity source in case of an overload or short circuit. The disconnect systems include circuit breakers and battery isolation devices. The electrical system design should:

- Provide high voltage, disconnecting contactors to isolate batteries in an emergency. These contactors should be located as close as possible to the "+" positive and "–" negative output terminals of the energy storage system.

3.1.6 On-Board Charging

The potential for electric shock also exists during battery recharging. The on-board charging system design should comply with UL-2202 and take into consideration the following issues. Transit agencies should consider the following design options. The charger should:

- Communicate with vehicle systems or energy storage device to sense the temperature, state of charge and/or voltage of the battery pack and follow an automatic charging protocol as suggested by the battery manufacturer.

- Shut off charging the batteries when the pack has attained the full charge state, or when an anomalous condition, such as high temperature or a fire condition, has been sensed.

- Be located and attached to the vehicle structure in such a way as to be electrically isolated from the vehicle chassis.

External charging requirements are discussed in the next chapter.

3.1.7 Power Train and Control Systems

Thermal degradation of the vehicle systems may also lead to failure of the components and a resulting loss of electrical isolation.

- Overheat detection system should be provided for the motor and the controller system.

- Adequate ventilation and cooling should be provided for the power train and prime mover motor(s) to ensure that overheating of the equipment does not result during normal operation on peak operation conditions.

- When the parking brake is applied, the direction control should automatically shift to neutral to prevent excess heat buildup.

3.2 HYDROGEN GASSING

Another safety issue of great importance is related to the charging of batteries. When some batteries are charged, the evolution of hydrogen occurs due to the dissociation of water in the electrolyte (especially in flooded lead-acid and NiCd batteries). Even sealed batteries, such as NiMH can discharge hydrogen due to overcharging. Prevention of potentially explosive reaction between hydrogen and oxygen gases is of paramount importance; this is achieved by performing battery charging operations in well ventilated areas to ensure that hydrogen concentrations remain below the lower explosion limit. Additionally, maintenance personnel should ensure that no ignition sources are near the battery charging stations. Use of listed battery charger/battery pack combinations that have been evaluated using SAE J1718 test method and identified as suitable for charging indoors is recommended.

- The battery housing/compartment should be designed with proper ventilation and meet the requirements of SAE J1766. The design should also ensure that moisture; road salt, rainwater, sand and soil, and other foreign debris cannot enter or accumulate within the battery compartment.

- Maintenance requirements for batteries should be clearly defined. In particular, the water level in flooded batteries should be maintained at the manufacturer suggested level to avoid "dry cell" conditions. Dry cell conditions can lead to formation of explosive gases during charge and discharge and pose a potential source of ignition and explosion.

- Active ventilation should be provided during the charging of the batteries so that flammable and explosive gases do not accumulate within the compartment.

Or

- Batteries should comply with the requirements of SAE J1718 for charging in enclosed spaces without vent fans.

- The vehicle manufacturer has certified concentrations of explosive gases, such as hydrogen in the battery box do not exceed 25 percent of the lower explosive limit during and following normal or abnormal charging and operation of the vehicle.

3.3 FIRE/TOXIC FUMES

In an electric bus or a hybrid-electric bus a significant amount of energy is stored, which creates the potential for energy to be released accidentally by a short circuit. While the short circuit carries electric risk, there also exists a potential for creating a high energy arc leading to the ignition of combustibles. Battery components, other than the plastic casing, are less susceptible to burning in a fire. However, many battery components (electrodes and electrolytes) are made of substances that when heated give off toxic fumes. In addition, several metals used in some batteries, lithium in particular, have the potential to burn at very high temperatures when ignited. Electrical isolation of the energy-storage system before any maintenance occurs on the batteries, cables and circuits is essential. When battery packs are removed or replaced significant care needs to be exercised to ensure that electrical systems are isolated and that no electrolyte leak occurs within the battery compartment. Additional inspections should make certain that no pathway for arcing is possible due to collection of dust or other contaminants.

A fire created by an electrical short circuit cannot be extinguished until the source of electrical energy is disconnected. A fire retardant barrier or coating between the batteries and battery box or the bus itself should be used to prevent, or at the very least delay, the spread of fire. The next level of safety would involve the installation of a fire suppression system to reduce the risk of the fire from spreading to other parts of the vehicle. Fires supression is not currently a requirement on diesel fueled buses, nor would it suppress an electrical short circuit arc.

It is recommended that transit agencies consider specifying that the vehicle have:

- An independently powered system of active thermal detection in the battery compartment that alerts the driver and/or personnel when the temperature is greater than 180°F;

- Battery box materials that are compatible and non-reactive with the battery electrolytes;

- Non-conductive battery box, or one coated with non-conductive materials;

- Battery modules should have terminals that are accessible and located facing towards the compartment opening.

- Battery modules properly secured to withstand road vibrations and designed to ensure that battery terminals do not come in contact with any part of the bus body or battery box and are not ejected, or leak, even under severe crash conditions.

- Battery overheat, fire or smoke conditions in the battery compartment should actuate a visual alarm at the operator's control panel. The specific type of alert should be indicated to the operator. An interlock to battery charging system should disable any charging underway in these conditions.

3.4 ELECTROLYTE (ACID) SPILLS

SAE and existing federal regulations require that batteries be designed to minimize the amount of battery electrolyte that could be spilled during a collision. Other safety measures transit agencies should consider:

- Require Material Safety Data Sheets be supplied for batteries

- Developing recommended response procedures for an electrolyte spill.

- Coordinating with, and developing appropriate emergency response materials for staff and first responders for neutralizing spills and treating chemical burns.

- Specifying that the battery compartment be designed to prevent all battery fluids from entering the passenger compartment during a vehicle crash.

3.5 VEHICLE (ELECTRICAL SYSTEM) MAINTENANCE ISSUES RELATED TO SAFETY

Maintenance of the electrical and traction systems in an electric or hybrid-electric vehicle is an extremely important function, which ensures not only the availability of the vehicle, but also its safe operation. Different items require different intervals of maintenance schedule. The inspection and maintenance protocol for the electrical and associated systems should include:

- Conducting periodic inspection of the battery systems to check the condition of the terminals, terminal connections and for moisture accumulation, if any, in the compartment. Also, a check for salt or other types of corrosion both on the terminals and on the compartment walls should be completed. A layer of salt or salt water on a normally non-conductive surface can provide a current path. If any condition that is out of the ordinary is found, immediate action should be taken to remedy the condition. Although batteries should be properly protected from moisture, dust and salt, inspection procedures should assume that it is possible for their infiltration.

- Ensuring that the flooded battery water level is within specified limits or specify batteries that do not require periodic watering.

- Periodically cleaning the battery packs when using any type of flooded battery systems so as to remove accumulation of electrolytes and salts created during charging. A build-up of salts can cause electrical shorts. Use a cleaning material that is not conductive and does not leave a residue.

- Ensuring that under no circumstances do the battery terminals and battery compartment lid touch if the battery compartment is made from conductive materials, such as aluminum.

- Regularly performing electric isolation tests (ground fault test). See SAE J1766 for details.

- Checking to ensure that the system monitoring of the battery temperature is working.

- Inspecting to ensure that the battery cables are not damaged or that the terminal connectors are not cracked. Check wire insulation for cuts, nicks or splits. Replace any defective cables found, which means NOT wrapping cables and wiring with electrical tape.

CHAPTER 4: Safety Issues in the Maintenance/Storage Facility

The intent of this section is to outline potential facility safety risks, and provide guidance on managing and mitigating them. Information is derived from existing codes and standards, industry input, and operational experience. The information is presented as a recommend action.

The general construction standards applicable to the bus maintenance/storage facilities (e.g., National Fire Protection Association (NFPA) 88A for vehicle storage and NFPA 30A for repair garages)[8] should be followed for the design and/or construction of buildings in which electric or hybrid-electric buses are parked and serviced. In addition, because of the prevalence of high voltage electric cables, wires, outlets and high power battery chargers in the garages servicing electric bus and hybrid-electric bus, applicable sections of the National Electric Code (NFPA 70) should be followed. Extra precaution should be taken in handling, storing and off-board charging of battery packs. Some of the related safety issues are indicated in the next sections. It is important that transit agencies adhere to the most current applicable codes and standards to ensure safety.

4.1 BATTERY OFF-LOADING & HANDLING

Battery packs in electric buses and hybrid-electric buses are bulky and typically weigh 600 – 2,000 pounds. Removing and reinstalling packs has to be conducted with care and precision. In performing these functions, the following safety issues need to be observed:

- Manually locking out of master disconnect.

- Ensuring that all battery box connections have been disconnected electrically from the bus systems. This can be enforced through physical design, such as tie down bolts located under cable leads, which can only be accessed after terminals have been disconnected.

- When battery packs are taken out of the bus, capping the terminals with an electrical insulating material to ensure that there is no possibility of shorting the terminals (e.g., contact with conductive parts). This can be imposed through physical design, such as female connectors on tub side with recessed conductor.

- Battery packs should be removed with the aid of mechanical devices such as a properly sized manual hoist, powered gantry crane, forklift, or a pallet jack. Irrespective of what mechanical device is used, considerations should be given to personnel safety. Consideration of the removal and replacement process should be part of the design of the battery packs. For example, battery pack designs should include lifting points to ensure safe removal and the vehicle should have adequate clearance space for its removal. Care should be taken to ensure that the compartment or any cables or connectors are not pinched, crushed, or damaged.

---

[8] More information about the NFPA and NFPA codes and standards can be found in Appendix A.

- If the battery pack consists of flooded acid batteries, ***great care*** should be exercised during the removal of the pack from the bus to ensure that no chemical spill occurs. Procedures should include maintaining the battery pack in a relatively horizontal position.

- Adequate and complete safety equipment and personal protective equipment should be provided. This includes insulted gloves and clothing, work and kneeling pads that are secure and properly insulated, and high-voltage rescue equipment. A good reference for this equipments and procedures are the maintenance personnel who service light and heavy rail equipment.

- If the battery pack is roof mounted proper personnel safety devices should be used. This may include safety harnesses and safety cables. Secure ladders or platforms should also be provided.

4.2 BATTERY STORAGE

- The storage racks to hold and store the battery packs should be structurally designed to hold the specified weight.

- Building storage locations of batteries must be ventilated to ensure that the gases emanating from the cells when the batteries are charged are purged out of the building quickly and efficiently. Reliance on SAE J1718 alone may not be sufficient for multiple packs charging simultaneously in the same area. In addition, the ventilation system should ensure that the concentrations of out-gas within the bus storage and repair buildings does not exceed the limit values specified under OSHA regulations. Blow out building panels are also recommended to prevent building damage, where appropriate.

- Battery storage during off-board charging should be such that the battery packs do not hinder the movement of the buses in the garage.

- Battery modules should be cleaned to ensure that there is no current leakage between the battery terminals during storage.

- Battery storage racks should be properly grounded so that personnel do not inadvertently become a ground path.

4.3 BATTERY CHARGING

- The battery charger should meet the requirements of SAE J1772 or J1773.

- Battery packs should be charged with equipment suitable for and compatible with the type and characteristics of the batteries, which means utilizing battery manufacturer approved chargers and adhering to the charging protocol. Because the charger will be connected to the building AC supply, the charger, whether on-board or off-board, should comply with UL2202 (per NEC Article 625). The UL 2231 personnel protection systems standards include guidance for protection against electric shock, evaluation of hardware and software controls and solid-state circuitry.

- The charger should be able to sense the battery pack's state of charge and identify when "end of charge" conditions are met and switch off or into a mode that does not overcharge the battery.

- The bus should not be MOVABLE when the charging door is open. It is also preferable to detect an open charging door than to check for current at the charging port, since the cable may be attached, but not energized.

- There should be indicators at the charging port that specify the bus is charging normally, has completed charging or has experienced a charging failure.

- Any fire detection and suppression systems placed close to the battery charging areas system must be compatible with electrical fires.

4.3.1 Battery Charger Safety and Location

Transit agencies should discuss their charger choice with their electricity provider to ensure that the type of electricity is available. Level 3 charging often requires wiring upgrades. Level 2 and Level 3 (typically fast charging) require the elements listed in Table 4.1.

Table 4.1: Battery Charger Requirements

	Voltage (VAC)	Current (amps)	Power (kVA)	Frequency (Hertz)	Phase
Level 2	208/240	32	6.7/7.7	60	Single
Level 3	208/240 or 480	400	192	60	Three

Source: Massachusetts Division of Energy Resources, 2000

Battery chargers should:

- Be capable of an automatic charging protocol (specified by the battery manufacturer) depending upon the state of charge condition of the batteries in the pack.

- Shut off, automatically, when the battery pack attains the rated full charge, in case of over temperature/fire conditions or with an occurrence of an anomalous condition.

- Include charging cables properly rated (for maximum voltage and current that can occur during charging) and of length less than 25 feet.

- Chargers should automatically detect a ground fault condition between the high and low voltage systems. The vehicle chassis should also be grounded while charging.

- Comply with UL 2202, UL 2251 and personnel protection systems covered by UL 2231-1, based on the requirements of the NEC Articles 511 and 625

The location of battery charging equipment should be taken into consideration. For example, if the charger is exposed to weather, it should be UL listed for outdoor use. The charging cables, wiring, materials should be sunlight (UV) resistant.

Charging equipment can produce considerable heat during the charging process. This heat must be dissipated by water or air-cooling of the equipment. Sufficient safeguards must be engineered into the charging units that if the temperature exceeds a set safe limit the charging is automatically shut off.

4.4 FACILITY FIRE DETECTION AND PROTECTION SYSTEMS

The fire protection system provided in the bus garage should be commensurate with the electrical charging activity and battery storage that occurs in the maintenance facility. NFPA 72 provides some guidance in this area. At the very minimum, the facility fire protection system should include the following:

- Smoke and heat detection systems near the charging units. The detection systems should provide both audible and visual alarms to personnel in the facility.

- Heat detection systems near battery storage areas, if they are separate from the charging area.

- Fire suppression system consisting of foam or fire extinguishing agents appropriate for Class C fires near the battery charging equipment.

- Water sprinkler based fire suppression system in other maintenance areas where there is no possibility of electrical fires.

CHAPTER 5: Personnel Training

Training of transit personnel involved in all levels of operation of electric or hybrid-electric buses is an important part in achieving a safe, economic and successful bus operation. Because it will require additional effort initially, transit agencies should address concerns from personnel and educate them as to why electric and hybrid-electric vehicles are an important element of the transit bus fleet. Personnel can influence future purchases if initial experiences are not positive, which can occur without proper training and support.

Additionally, transit agencies should be aware that new vehicle technology will require an investment of time, particularly in trouble shooting tasks, and in securing additional unique replacement parts. Transit agencies with electric and hybrid-electric technology, such as New York City (NYC) highlighted on the next page, reported that personnel experienced a learning curve. Hence, the more experience and training provided in the beginning, whether attending classes or shadowing personnel at a facility with the technology, the better equipped transit agency personnel will be in implementing a smooth transition.

While electric and hybrid-electric buses are driven and operated similarly to conventional buses, transit agencies are still encouraged to allow for one week minimum for personnel training upon the arrival of new electric and hybrid-electric buses. In addition to training its own personnel, a transit system should provide information and some basic training to local emergency response personnel who may be required to respond to accidents involving buses with high voltage systems.

Some of the training issues that a transit system should consider implementing are discussed below.

5.1 TRAINING OF TRANSIT VEHICLE OPERATORS

The efficiency of an electric bus depends significantly upon the extent of training received by bus drivers on the proper operation of the electric traction system and its details, such as regenerative braking. The training of an operator should consist of at least, the following elements:

- Overview of the dashboard controls and warning signals, and what the correct procedure is when one or more warning lights or alarms are initiated.

- Concepts, working principles and details of regenerative braking, mechanical braking, hill holding and roll back. Also, the details of regenerative braking systems and how they differ from the conventional brake and retarders.

- Optimal levels of acceleration and deceleration to maximize the efficiency of the vehicle.

- Items to perform in the case of "low power battery" indication. Also, how to shut down the bus safely both after a normal operation, and in an emergency.

- Specific safety issues related to batteries. This should include electrocution hazards due to the high voltage, arcing, degassing during battery recharging and fire in the case of short circuit.

- Locations of various emergency cut out switches to disconnect and disable the electrical system in the bus.

- Actions to take or avoid in an emergency, such as advising emergency responders about the concerns of potential hazardous materials from the batteries and high voltage elements.

- Availability of proper rescue and fire equipment.

- Inspection of battery management system to ensure that batteries are operating within parameters and inspection per manufacturer's recommendation.

- How to conduct a proper daily walk around inspection before starting revenue service.

5.2 TRAINING OF MAINTENANCE PERSONNEL

The training of maintenance personnel should include:

- Significant emphasis on the dangers of high voltage and the differences between the electrical system in a hybrid-electric bus or an electric bus and in conventional buses.

- The means, such as instruments and checklists, by which maintenance personnel and supervisors ensure that any system or component of an electric bus or hybrid-electric bus is declared safe to inspect, open, modify, replace or maintain.

- An overview of the electric bus systems and items that can be maintained in the shop.

- System logic, circuits, and how to conduct a fault diagnosis. This should include the current paths, grounding principles, interlocks and logic conflicts.

- Diagnostic equipment training.

NYC Transit Case Study

A Department of Energy transit bus evaluation project collected operating costs, efficiency, emissions, and overall performance of ten heavy-duty low-floor hybrid buses and fourteen conventional high-floor diesel transit buses operated by New York City Transit under similar duty cycles. These pre-production buses met the project's design goals for low emissions, improved fuel economy and performance equal or better than standard buses.

Some conclusions from the project included:

- Facility conversion for accommodating hybrid buses was minor compared to preparing for CNG vehicles.

- The hybrid buses required extra time for servicing than conventional buses to troubleshoot and fix problems.

- Maintenance costs for hybrids were 76% - 150% higher than conventional buses.

- There is a strong indication that the maintenance cost differences between hybrid and conventional diesels will fall significantly for the next generation of vehicles.

Source: Battelle, July 2002.

- Battery issues including the different types of batteries, their chemistry and their characteristics.

- Battery charging regimes, equalizing charges in cells, over and undercharging and their consequences.

- Battery vent and watering systems and inspecting battery fluid levels, where appropriate.

- Electrical safety and items to consider before disconnecting and connecting electrical circuits, load testing, testing of battery capacity and efficiency.

- Removal from and reinstallation of battery packs into the bus compartment.

- Working with caustic materials, acids and how to neutralize battery fluid spills.

- Sensitivity issues for power and/or automatic vehicle washing, which may push water into vehicle.

- Out-gassing from batteries, flammability and toxicity of gases emanating from batteries.

- Proper and environmentally acceptable disposal or recycling of batteries, battery fluids, and other wastes.

- Performing periodic inspection of critical items and initiating preventive maintenance versus reactive maintenance.

- A systematic approach to documenting all findings during an inspection and prior to the performance of a maintenance procedure. This approach should include descriptions of battery terminal conditions, electrical isolation measurement of each battery box, battery box temperatures, evidence of battery module motion relative to the battery box, battery box attachment condition, and condition of propulsion wiring and connectors.

5.3 EMERGENCY RESPONSE PERSONNEL TRAINING

Outwardly, an electric or hybrid-electric bus may not look any different from other buses. However, fire characteristics and the techniques of fighting electrical fires are different from those that occur when responding to a combustibles fire. Therefore, it is essential that the emergency response personnel such as fire fighters be provided with information about the characteristics and design differences between electric and hybrid-electric buses, and conventional buses.

The training and information provided to the emergency response personnel should include:

- How to distinguish an electric bus from other bus types. This may include the location and type of decals on the bus, and the absence of engine noise or combustion gas exhaust vent in the case of pure electric vehicle.

- Name and telephone number of an emergency contact person within the transit system to reach in case of an emergency.

- An overview of the principles of operation, important components (batteries, motor, controller panel, inverter and on-board charger) and their locations in an electric or hybrid-electric bus.

- In the case of a hybrid bus, the type of the auxiliary power unit (internal combustion engine or the fuel cell) also should be identified. Also, the different types of fuels used to power the auxiliary power units (diesel, gasoline, CNG, methanol and hydrogen) should be indicated along with their properties. The locations of on-board fuel storage tanks, fire fighting techniques and acceptable extinguishing agents should be explained.

- Location of emergency cut out switches to disconnect the electrical system from the energy storage devices. Proper procedures for disconnecting batteries.

- Characteristics of electrical fires and methods to deal with them. This includes disconnecting electrical supply for such equipment as motors, inverters and charging units. In the case of batteries, isolating the batteries and removing them from the body of the bus if feasible or in-situ quenching of the fire using suitable extinguishing agents.

- Electrical shock hazards when water is sprayed on live batteries or electrically energized systems in the bus. Municipal water has dissolved salts and this makes the water electrically conductive. Emergency service personnel directing water on to an electrically active part of the system may experience a life threatening high voltage electric shock.

- Types of batteries used in the electric and hybrid-electric buses and their chemistry. Also how to treat chemical burns and neutralize spilled battery fluids. Responders should also be made aware of any potential explosive or toxic gas hazards the batteries may pose.

CHAPTER 6: REFERENCES

Alternative Fuels Data Center, "Alternative Fuels," Department of Energy, 2002, website available at http://www.afdc.doe.gov/altfuels.html

Chandler, Kevin, Kevin Walkowicz and Leslie Eudy, *New York City Transit Diesel Hybrid-Electric Buses: Final Results*, U.S. Department of Energy, July 2002. Available at www.afdc.doe.gov

Electric Power Research Institute, "Emergency Response to Electric and Hybrid-Electric Buses," Addendum to the "Emergency Response to Electric Vehicles" April 2001.

Electric Vehicle Association of Americas, "Electric Vehicle Association of Americas," 2002, website at http:// www.evaa.org

Federal Transit Administration (FTA), Four Year Report on Battery-Electric Transit Vehicle Operation at the Santa Barbara Metropolitan Transit District, FTA-CA-26-0019-95-1, US DOT, 1995.

Massachusetts Division of Energy Resources, "Installation Guide for Electric Vehicle Charging Equipment," September 2000. Available at www.state.ma.us/doer/programs/ev/charger1.pdf

Northeast Sustainable Energy Association, *2002 Tour de Sol Rules*, Version 14.02, December 15, 2001.

National Research Council, "Effectiveness of the United States Advanced Battery Consortium as a Government-Industry Partnership," Item 4: Technical Progress, p 33-46, *Report of the Committee to Review the U.S. Advanced Battery Consortium Electric Vehicle Battery Research and Development Project Selection Process*, National Academy Press, 1998. Available at http://search.nap.edu/nap-cgi/napsearch.cgi

Transit Cooperative Research Program, *Hybrid-Electric Transit Buses: Status, Issues, and Benefits*, Report 59, Transportation Research Board, 2000.

Underwriters Laboratory Inc., "UL and Electric Vehicles," 2002, available at http://www.ul.com/ulev/

CHAPTER 7: APPENDIX A: APPLICABLE REGULATIONS, CODES, STANDARDS & RESOURCES

The following are applicable regulations related to the design and operation of an electric bus or a hybrid-electric bus and the associated storage/maintenance facility.

REGULATIONS

- FTA Regulation on Bus Testing in 49 CFR, Part 665.
- NHTSA Federal Motor Vehicle Safety Standards in 49 CFR, Part 571.
- State regulations

CODES & STANDARDS

NATIONAL FIRE PROTECTION ASSOCIATION Contact: P.O. Box 9101, Quincy, MA 02269-9101; Customer Service: 1.800.344.3555; Website: www.nfpa.org

NFPA is an independent, nonprofit organization whose mission is to reduce the worldwide burden of fire and other hazards on the quality of life by providing and advocating scientifically based consensus codes and standards, research, training, and education.

- NFPA 70: National Electrical Code, 2000 edition.
- NFPA 72: National Fire Alarm Code. 2002 edition
- NFPA 88A: Standard for Parking Structures, 1998 edition.
- NFPA 30A: Standard for Motor Fuel Dispensing Facilities and Repair Garages, 2000 edition.

SOCIETY OF AUTOMOTIVE ENGINEERS Contact: 400 Commonwealth Drive, Warrendale, PA 15096-0001; Customer Service: 724.776.0790; Website: www.sae.org

SAE is a technical information resource and provides expertise in designing, building, maintaining, and operating self-propelled vehicles for use on land or sea, in air or space.

- SAE Recommended Practice J1673: High Voltage Automotive Wiring Assembly Design

- SAE Recommended Practice J1718: Measurement of Hydrogen Gas Emission From Battery-Powered Passenger Cars and Light Trucks During Battery Charging

- SAE Recommended Practice J1742: Connections for High Voltage On-Board Road Vehicle Electrical Wiring Harnesses.

- SAE Recommended Practice J1766: Recommended Practice for Electric and Hybrid Electric Vehicle Battery System Crash Integrity Testing

- SAE Recommended Practice J1797: Packaging of Electric Vehicle Battery Modules

- SAE Recommended Practice J1798: Performance Rating of Electric Vehicle Battery Modules

- SAE Recommended Practice J2344: Guidelines for Electric Vehicle Safety.

- SAE Recommended Practice J2293: Energy Transfer System for Electric Vehicles

INTERNATIONAL ORGANIZATION FOR STANDARDIZATION Contact: 1, rue de Varembé, Case postale 56 CH-1211 Geneva 20, Switzerland; Customer Service + 41 22 749 01 11; Website: www.iso.org

The ISO is a worldwide federation of national standards bodies from more than 140 countries. ISO is a non-governmental organization with a mission to promote the development of standardization and related activities in the world with a view to facilitating the international exchange of goods and services, and to developing cooperation in the spheres of intellectual, scientific, technological and economic activity.

- ISO Standard 6469 parts 1, 2, and 3: International Guidelines for wiring, safety issues and electrical isolation.

Electric Power Research Institute Contact: 3412 Hillview Avenue, Palo Alto, California 94304; Website: www.epri.com

EPRI is a non-profit energy research consortium for the benefit of utility members, their customers, and society.

- EPRI's Electric Bus Technical Specifications

UNDERWRITERS LABORATORIES, INC. 333 Pfingsten Road, Northbrook, IL 60062-2096. Customer Service: 708.272.8800; Website: www.ul.com

Underwriters Laboratories Inc. (UL) is an independent, not-for-profit product safety testing and certification organization.

- UL 50: Standard for Enclosures for Electrical Equipment

- UL 991: Standard for Tests for Safety-Related Controls Employing Solid-State Devices

- UL 1244: Electrical and Electronic Measuring and Testing Equipment
- UL 1439: Determination of Sharpness of Edges on Equipment
- UL 1998: Standard for Safety-Related Software
- UL 2202: Electric Bus Charging System Equipment
- UL 2231: Personnel Protection Systems for Electric Bus Charging Circuits
- UL 2251: Plugs, Receptacles, and Couplers for Electric Vehicles

RESOURCES

UNITED STATES ADVANCED BATTERY CONSORTIUM, part of the United States Council for Automotive Research. Website:
http://www.uscar.org/consortia&teams/consortiahomepages/con-usabc.htm

VOLPE NATIONAL TRANSPORTATION SYSTEMS CENTER, 55 Broadway, Kendall Square, Cambridge, MA 02142

"Design Guidelines for Bus Transit Systems Using Compressed Natural Gas As An Alternative Fuel", Technology and Management Systems, 1996. (FTA-MA-26-7021-96-1)

"Design Guidelines for Bus Transit Systems Using Alcohol Fuel (Methanol and Ethanol) as an Alternative Fuel", Technology and Management Systems, 1996. (FTA-MA-26-7021-96-3)

"Design Guidelines for Bus Transit Systems Using Liquefied Petroleum Gas (LPG) as an Alternative Fuel", Technology and Management Systems, 1996. (FTA-MA-26-7021-96-4)

"Design Guidelines for Bus Transit Systems Using Liquefied Natural Gas (LPG) as an Alternative Fuel", Technology and Management Systems, 1997. (FTA-MA-26-7021-97-1)

"Design Guidelines for Bus Transit Systems Using Hydrogen as an Alternative Fuel", Technology and Management Systems, 1999. (FTA-MA-26-7021-98-1)

These reports are available for free at http://transit-safety.volpe.dot.gov/Publications